YOUR KNOWLEDGE HAS VALUE

- We will publish your bachelor's and
  master's thesis, essays and papers

- Your own eBook and book -
  sold worldwide in all relevant shops

- Earn money with each sale

Upload your text at www.GRIN.com
and publish for free

Bibliographic information published by the German National Library:

The German National Library lists this publication in the National Bibliography; detailed bibliographic data are available on the Internet at http://dnb.dnb.de .

Imprint:

Copyright © 2017 GRIN Verlag, Open Publishing GmbH
Print and binding: Books on Demand GmbH, Norderstedt Germany
ISBN: 9783668481145

This book at GRIN:

http://www.grin.com/en/e-book/370209/calculator-tutor-for-casio-fx-991es-plus-for-the-use-in-senior-high-schools

Farouq Sessah Mensah

Calculator Tutor for Casio fx-991ES/Plus. For the use in Senior High Schools

GRIN Publishing

GRIN - Your knowledge has value

Since its foundation in 1998, GRIN has specialized in publishing academic texts by students, college teachers and other academics as e-book and printed book. The website www.grin.com is an ideal platform for presenting term papers, final papers, scientific essays, dissertations and specialist books.

Visit us on the internet:

http://www.grin.com/

http://www.facebook.com/grincom

http://www.twitter.com/grin_com

CALCULATOR TUTOR

FOR

SENIOR HIGH SCHOOLS

By

Farouq Sessah Mensah

DEDICATION

This book is dedicated to my parents, Mr. Kofi Ibrahim Mensah and Mrs. Sakeena Mensah, whose love and unfailing support has made me what I am. I love you Dad and Mum. It is also dedicated to my Uncle Ahaji Ismail Adam of Bank of Ghana.

ACKNOWLEDGEMENT

My ultimate appreciation goes to the Almighty Allah for giving me the wisdom, knowledge and strength to undertake this work.

My thanks go to Musah Kofi Amissah (Wakito), ICT Tutor, Ekumfi T. I. Ahmadiyya SHS whose advice and motivation propelled me into writing this book. I am also grateful to Mr. Jebreel Odoom of Ekumfi T. I. Ahmadiyya SHS (Formerly of Adisadel Collage) who read through the manuscript and suggested improvement and to Nana Esi Bedua Taylor for her encouragement.

I am also grateful to numerous colleagues and friends who made useful criticisms and comments during the writing of the manual. I am especially grateful to Obed Yankson (Mathematics Tutor, Kwagyery Aggrey SHS), Issach Egyir (Head of Mathematics Department, Ekumfi T. I. Ahmadiyya SHS), Usman Kofi Mensah (Ezi Savings and Loans), George Bello, (Coordinator, CCE, UCC, Zenith Center), Edward Kweku Walker (Mathematics Tutor, University JHS, UCC)   Sarah Ansah and Muzafar Ackonu for excellent typing work.

Finally, my thanks goes to all teachers of Ekumfi T. I. Ahmadiyya SHS for their cooperation during my period of teaching and to my numerous Students whose effort led to the writing of this manual. However, any shortcomings the book may possess are entirely my own responsibility.

ABOUT THE AUTHOR

Farouq Sessah Mensah was born on the March, 1988 at Akim Oda in the Eastern Region of Ghana. He started is basic education at Flagstaff House Basic School and completed in April, 2003.

Farouq then continued his secondary education at T.I. Ahmadiyya SHS, Kumasi in the same year. He studied Business and completed in May, 2006.

After, secondary education He continued to the University of Cape Coast where He read Bachelor of Education (Mathematics) and completed in May 2011.

Had his national service at University Junior High School, UCC where he taught Mathematics. He is currently teaching Mathematics at Ekumfi T.I. Ahmadiyya SHS.

The Author's experience in teaching Mathematics propelled him to write this book to aid students write Mathematics papers with ease, faster and accurately.

PREFACE

The Calculator Tutor is written with the core objective to assist students preparing for West African Secondary Certificate Examination (WASSCE) and for students in the tertiary institutions to easily make good use of the calculator to solve most of the objective questions and to verify their answers for the written papers.

The book has been structured to alley students' fear of answering some complicated objective questions in both Core and elective Mathematics papers. Model questions and step by step ways of solving them with the scientific calculator have been provided to facilitate grasping of the content.

The author is optimistic that any student who constantly practice the content of this book will not be found wanting as far as mathematics examination is concerned. It must however, be noted that I am not a master of all knowledge. Consequently, I warmly welcome any useful criticism and suggestions for future work.

TABLE OF CONTENTS

GETTING STARTED

FEATURES ON THE CALCULATOR AND THEIR FUNCTIONS

Before starting a calculation, you must first enter the correct mode pertaining to the type calculation you want to make. The mode are indicated in the table below.

Mode Description	Key Operation	Notation
Basic Arithmetic	[MODE] [1]	COMP
Complex Number	[MODE] [2]	CMPLX
Statistical Calculation	[MODE] [3]	REG
Base-N Calculation	[MODE] [4]	BASE
Equation	[MODE] [5]	EQN
Matrix Calculation	[MODE] [6]	MAT
Table Calculation	[MODE] [7]	TABLE
Vector Calculation	[MODE] [8]	VCT
Decimal Place	[SHIFT] [MODE] [6]	Fix
Significant Figures	[SHIFT] [MODE] [7]	Sci

****NB: There are other modes for other calculations.***

MODE INITIALIZATION

The calculation mode can be returned to the initial default by pressing the **[SHIFT] [CLR] [2] [MODE] [=] [AC]** or simply **[MODE] [1] (COMP)**.

All data currently in the calculator memory is cleared when this key sequence is used: **[SHIFT] [CLR] [3][=] [AC]** .

STORE AND RECALL

To store number or answer on the calculator memory, the following procedure must be followed

- ➢ Punch the number to be stored and press **[=]**;
- ➢ Press **[SHIFT] [STO]** followed by the alphabet you want to store the number.

The stored number could be recalled by pressing **[RCL]** followed by the alphabet that the number was stored.

Example

To store the number **2014** unto the calculator memory, the following key sequence must be followed:

- ➢ Enter **2014 [=] [SHIFT] [STO] [A]**. The number has been stored in **A**.

To retrieve the stored number, the following key sequence must be followed:

- ➢ Press **[RCL] [A]**.

EQUATIONS

For the calculator to be used to do calculations on equations, the **EQN (equation mode)** should be turned on. Thus **[MODE] [5]** will turn the **EQN** on. The equation function is used for the calculation of simultaneous, quadratic and cubic equations.

SIMULTANEOUS EQUATION (EQUATIONS IN TWO/THREE VARIABLES)

To turn on the simultaneous equation function: **[MODE] [5]**. Option **1** and **2** are the simultaneous equation modes. Select option 1 if the equation has two unknown variables or option 2 if it has three unknown variables. Here we are interested in two unknown variables. The procedure is exactly the same for three unknown variables.

1: **[anX + bnY = cn]** (Simultaneous Equation In Two Variables)

2: **[anX + bnY + CnZ = dn]** (Simultaneous Equation In Three Variables)

Example 1

Find the value of x and y in the following equations: $2x + y = 5$ and $2x + 2y = 6$

a1: Coefficient of x in equation 1 $= 2$

b1: Coefficient of y in equation 1 $= 1$

c1: The constant in equation 1 = 5

a2: Coefficient of x in equation 2 = 2

b2: Coefficient of y in equation 2 = 2

c2: The constant in equation 2 = 6

Procedure

[MODE] [5] [1] (since the equation are in two variables)

	a	b	c
1				(SCREEN DISPLAY)
2				

➢ Row 1: [ENTER] 2 [=] 1 [=] 5 [=]

➢ Row 2: [ENTER] 2 [=] 2 [=] 6 [=]

Result: $(x = 2$ [=] y = 1) (Bravo!).

CHECK!

Putting $x = 2$ and $y = 1$ into the equations:

$2x + y = 5$ and $2x + 2y = 6$

$2(2) + 1 = 5$ and $2(2) + 2(1) = 6$ **(Interesting!)**

Example 2

Solve the equation $3x - y = 7$ and $x + y = 5$

Procedure

➢ [MODE] [5] [1]

➢ Row 1: [ENTER] 3 [=] − 1 [=] 7 [=]

➢ Row 2: [ENTER] 1 [=] 1[=] 5 [=]

Result: $(x = 3 \; [=] \; y = 2)$.

Example 3

Solve the equation $-2x + 3y = 19$ and $2x + y = 1$

Procedure

➢ [MODE] [5] [1]

➢ Row 1: [ENTER] − 2 [=] 3 [=] 19 [=]

➢ Row 2: [ENTER] 2 [=] 1[=] 1 [=]

Result: $(x = -2 \; [=] \; y = 5)$.

Example 4

Find the truth set of the following simultaneous equations: $3a -$ $2b = 8$ and $\frac{a}{3} + \frac{b}{2} = \frac{5}{4}$

Procedure

➢ [MODE] [5] [1]

➢ Row 1: [ENTER] 3 [=] − 2 [=] 8 [=]

➢ Row 2: [ENTER] $\frac{1}{3}$ [=] $\frac{1}{2}$[=] $\frac{5}{4}$ [=]

Result: $(a = 3 \,[=]\, b = 0.5)$. (***The result is given in terms of x and y so x = a and y = b ***)

Truth set $= \{a: a = 3 \text{ and } b: b = 0.5\}$

<u>Example 5</u>

Solve the set of equations

$$x + 2y + z = 4 \ldots \ldots \ldots \quad (1)$$

$$3x - 4y - 2z = 2 \ldots \ldots \quad (2)$$

$$5x + 3y + 5z = -1 \ldots \quad (3)$$

<u>Procedure</u>

➢ **[MODE] [5] [2]**
➢ Row 1: **[ENTER] 1 [=] 2 [=] 1 [=] 4 [=]**
➢ Row 2: **[ENTER] 3 [=] − 4 [=] − 2 [=] 2 [=]**
➢ Row 3: **[ENTER] 5 [=] 3 [=] 5 [=] − 1 [=]**

Result: $(x = 2 \,[=]\, y = 3 \,[=]\, z = -4)$

WHY NOT TRY THE FOLLOWING

1. $2x - 5y = -6$ and $4x - 3y = -12$
   Result $x = -3$ and $y = 0$

2. $a + b = -4$ and $4a + b = 2$

Result $a = 2$ and $b = -6$

3. Find the value of x, y and z if $5x - 2y + 3z = 16$; $2x + 3y - 5z = 2$ and $4x - 5y + 6z = 7$

Result $x = 3, y = 7$ and $z = 5$

QUADRATIC & CUBIC EQUATION

The key sequence for quadratic equation is given as [MODE] [5][3] . The option '3' is used for solving quadratic equations and option '4' is for Cubic Equations thus the option 3 and 4 are the polynomial modes- choose option 3 if you have a quadratic equation and option 4 if you have a cubic equation.

General procedure:

To solve a quadratic or cubic polynomial equation:

1. Go to **EQN** mode and select option 3 for quadratic equation. Thus, [MODE] [5][3]

2. Enter the polynomial coefficients, from highest degree downward.

   Each coefficient is followed by [=] in order to go the next coefficient

3. Press [=] again and you will be shown the roots of the equation.

4. Note that complex-valued solutions are included. Thus it will provide you with all roots, both real and complex.

5. Cubic equations $ax^3 + bx^2 + cx + d = 0$ are also solved the same way using the calculator.

The general equation of quadratic equation is given as $ax^2 + bx + c = 0$.

Let's use examples to illustrate how such equations are solved using your calculator.

Example 1

Find the roots of the equation $2x^2 - 6x + 4 = 0$

Procedure

➢ Press **[MODE] [5] [3]**

➢ Enter the coefficients in the equation ($a = 2, b = -6$ and $c = 4$)

➢ **[2] [=] [−6] [=] [4] [=]**

Result: $x_1 = 2$ [=] $x_2 = 1$

Example 2

Solve $3x^3 + 2x^2 + 4x + 5 = 0$

Procedure

> Press **[MODE] [5] [4]**
> Enter the coefficients in the equation ($a = 3$, $b = 2$, $c = 4\ and\ d = 5$)
> **[3] [=] [2] [=] [4] [=] [5] [=]**

Result: $x_1 = -1$ **[=]** $x_2 = \dfrac{1}{6} + 1.2802i$ **[=]** $x_3 = \dfrac{1}{6} - 1.2802i$

Example 3

Find the truth set of the equation $3x^2 - 8x + 5 = 0$

Procedure

> Press **[MODE] [5] [3]**
> **[3] [=] [−8] [=] [5] [=]**

Result: $x_1 = \dfrac{5}{3}$ **[=]** $x_2 = 1$

CHECK WITHOUT CALCULATOR!

$3x^2 - 8x + 5 = 0$

$3x^3 - 3x - 5x + 5 = 0$

$3x(x - 1) - 5(x - 1) = 0$

$(3x - 5)(x - 1) = 0$

$3x - 5 = 0$ or $x - 1 = 0$

$x = 5/3$ or $x = 1$

Example 4

Find the truth set of $x^3 - x^2 - 9x + 9 = 0$

Procedure

➢ Press **[MODE] [5] [4]**

➢ **[1] [=] [−1] [=] [−9] [=] [9] [=]**

Result: $x_1 = -3$ [=] $x_2 = 3$ [=] $x_3 = 1$

Example 5

Find the roots of the quadratic equation $5x^2 + 2x - 3 = 0$

Procedure

➢ Press **[MODE] [5] [3]**

➢ **[5] [=] [2] [=] [−3] [=]**

Result: $x_1 = \dfrac{3}{5}$ [=] $x_2 = -1$

 *TRY IT OUT!*

1. Solve the equation $x^3 + x^2 - 81x - 81 = 0$

Result: $x_1 = 9$ [=] $x_2 = -1$ [=] $x_3 = -9$

2. Find the roots of the quadratic equation $2x^3 + 5x^2 - 14x - 8 = 0$

   Result: $x_1 = -3$ [=] $x_2 = 3$ [=] $x_3 = 1$

3. Solve the equation $x^2 + 3x - 10 = 0$

   Result: $x_1 = 2$ [=] $x_2 = -5$

SIGNIFICANT FIGURES AND DECIMAL PLACES

Approximations are simply manipulated with the calculator when the number of significant figures or the decimal places are specified.

General Procedure

1. Specify the number of decimal places or significant places you want.
2. Punch the number unto the calculator and press [=]
3. The result would be displayed. For significant figures, the result would be displayed in standard form.
4. The key operation for significant figures and decimals places are [SHIFT] [MODE](Fix; Sci; Norm).
5. Where **Fix** is **Decimal Places** and **Sci** is **Significant Figures.**

Example 1

Correct 287530 to

a. 4 significant figures
b. 3 significant figures
c. 2 significant figures

Procedure

a. [SHIFT] [MODE] [7] and then specify the number of significant figures you want i.e. [4] 287530 [=] (*result* $2.875 \times 10^5 = 287500$)

b. [SHIFT] [MODE] [7] [3] 287530 [=] (*result*: $2.88 \times 10^5 = 288000$)

c. [SHIFT] [MODE] [7] [2] 287530 [=] (*result*: $2.9 \times 10^5 = 290000$)

Example 2

Correct each of the following to numbers to 2 significant figures

a. 0.0496

b. 0.0996

Procedure

[SHIFT] [MODE] [7] [2]

   a. 0.0496 [=] (result: $5.0 \times 10^{-2} = 0.050$)
   b. 0.0996 [=] (result: $1.0 \times 10^{-1} = 0.100$)

Example 3

Correct 23.4625 to

a. 1 decimal place

b. 2 decimal place

c. 3 decimal place

Procedure

a. [SHIFT] [MODE] [6] and then specify the number of significant figures you want i.e. [1] 23.4625 [=] (result: 23.5)

b. [SHIFT] [MODE] [6] [2] 23.4625 [=] (result: 23.46)

c. [SHIFT] [MODE] [6] [3] 23.4625 [=] (result: 23.463)

 *TRY IT OUT!*

a. Correct 0.7625 to 3 decimal place
    Results: 0.763

b. Correct 4.7625 to 3 significant figure
    Results: $4.76 \times 10^0 = 4.76$

ARITHMETIC

A. FRACTIONS

The calculator uses the popular order of operation 'BODMAS' in its calculations.

The questions are punched directly onto the screen; no 'mode' is required. The result can be converted between decimal and fraction (proper or mixed). When the result is in mixed number, it can be converted to improper fraction and vice versa by pressing [S ↔ D]. The fraction (proper or improper) can be entered on the calculator using the key [▬] and the mixed number by the key [SHIFT] [▬]

The use of parenthesis is very useful when dealing with algebra. The parenthesis tells which part of the algebra the calculator should tackle first.

Example 1

Perform the following operations

a. $3/4 + 7/8 + 1/2$

b. $3\frac{1}{2} + 2\frac{2}{3}$

Procedure

a. $[\frac{3}{4}]$ [+] $[\frac{7}{8}]$ [+] $[\frac{1}{2}]$ [=] (Result $\frac{17}{8}$). The result is in improper fraction form, to retrieve it in decimal form, just press [S $\leftrightarrow$ D]. (Result: 2.125)

b. $[3\frac{1}{2}]$ [+] $[2\frac{2}{3}]$ [=] (Result: $\frac{37}{6}$)

Example 2

Evaluate

a. $(\frac{3}{2} + 7)/(4\frac{1}{3} - 3)$

b. $\dfrac{^3/_2 + ^3/_4 + ^5/_4}{^7/_8 - ^1/_4}$

Procedure

a. $\left[\left([\frac{3}{2}] \; [+] \; 7\right)\right] \; [\div] \left[\left([4\frac{1}{3}] \; [-] \; 3\right)\right]$ [=] (Result: $\frac{51}{8}$)

b. $\left[\left([\frac{3}{2}] \; [+] \; [\frac{3}{4}] \; [+] \; [\frac{5}{4}]\right)\right] \; [\div] \; \left[\left([\frac{7}{8}] \; [-] \; [\frac{1}{4}]\right)\right]$ [=] (Result: $\frac{28}{5}$)

Note the parenthesis, which tells the calculator that the $(^7/_8 - ^1/_4)$ is dividing the whole of $(^3/_2 + ^3/_4 + ^5/_4)$. It is very important to know that*

Example 3

Arrange $^4/_5$, $^5/_6$, $^7/_{12}$, $^2/_3$ and $^3/_8$ in descending order.

Procedure

For such questions, use the calculator to convert each fraction to decimal and compare them.

$[\frac{4}{5}]$ [=] [S ↔ D] (Result: 0.8)

$[\frac{5}{6}]$ [=] [S ↔ D] (Result: 0.833333)

$[\frac{7}{12}]$ [=] [S ↔ D] (Result: 0.583333)

$[\frac{2}{3}]$ [=] [S ↔ D] (Result: 0.666667)

$[\frac{3}{8}]$ [=] [S ↔ D] (Result: 0.375)

Now in descending order $^5/_6$, $^4/_5$, $^2/_3$, $^7/_{12}$ and $^3/_8$

B. PERCENTAGES

The percentage sign can be called from the [shift] [( ].

Example 1

Express 12cm as a percentage of 60cm

Procedure

12 [÷] 60 [SHIFT] [%][=] (Result: 20)

Example 2

Express $50 as the percentage of $180?

Procedure

50 [÷] 180 [SHIFT] [%] [=] (Result: 27.778)

Example 3

Find 20% of 150 and then subtract the result from 150

Procedure

150 [×] 20 [SHIFT] [%] [=] (Result: 30)

150 − [ANS] [=] (Result: 120)

C. SURDS

You can use your calculator to solve surd questions. The procedure is straight forward: punch the expression onto the calculator screen and press [=].

Example 1

Simplify $\sqrt{500} - \sqrt{125}$

Procedure

$[\sqrt{500}]$ $[-]$ $[\sqrt{125}]$ $[=]$ (Result: $5\sqrt{5}$). The result is in the surd form, to retrieve it in decimal form, just press $[S \leftrightarrow D]$. (Result: 11.18033989)

Example 2

Simplify $\sqrt{98} + \sqrt{50} + \sqrt{18}$

Procedure

$[\sqrt{98}]$ $[+]$ $[\sqrt{50}]$ $[+]$ $[\sqrt{18}]$ $[=]$ (Result: $15\sqrt{2}$)

Example 3

Simplify $(3 - 2\sqrt{2})(3 + 2\sqrt{2})$

Procedure

$([3]$ $[-]$ $[2\sqrt{2}])$ $([3]$ $[+]$ $[2\sqrt{2}\,])$ $[=]$ (Results:1)

D. INDICES

Indices can be evaluated using the calculator. The key for exponent or index is $[x^{\blacksquare}]$.

Example 1

Without using tables, simplify the following:

a. $(\frac{64}{25})^{\frac{1}{2}}$

b. $(\frac{8}{27})^{-\frac{2}{3}}$

Procedure

a. $[(\frac{64}{25})^{\frac{1}{2}}] [=]$ (Result: $1.6 = \frac{8}{5}$)

b. $[(\frac{8}{27})^{-\frac{2}{3}}]] [=]$ (Result: $2.25 = \frac{9}{4}$)

Example 2

Simplify the following

a. $(27 \times 3^{-2}) (8 \times 2^{-3})$

b. $27^{\frac{2}{3}} \times 64^{\frac{1}{3}} \div 81^{\frac{1}{2}}$

Procedure

a. $([27] [\times] [3^{-2}]) ([8] [\times] [2^{-3}]) [=]$ (Result :3)

b. $[27^{\frac{2}{3}}] [\times] [64^{\frac{1}{3}}] [\div] [81^{\frac{1}{2}}] [=]$ (Result: 4)

Example 3

If $3^x = 9$ find the value of x

Procedure

[3] $\left[x\right]$ [ALPHA] [ )] [→] [ALPHA] [CALC] [9] [SHIFT]

[SOLVE] [=] (Result: $x = 2,\ L - R = 0$)

Example 4

Solve for x in the following equations

a. $3^{x-1} = 81$

b. $3^{2x-1} = \dfrac{1}{27}$

Procedure

(a) [3] [$x^{\blacksquare}$] [ALPHA] [ )] [−] [1] [→] [ALPHA] [CALC] [81]
   [SHIFT] [SOLVE] [=]   (Result: $x = 5,\ L - R = 0$)

(b) [3] [$x^{\blacksquare}$] [2] [ALPHA] [ )] [−] [1] [→] [ALPHA] [CALC] $[\frac{1}{27}]$
   [SHIFT] [SOLVE] [=]   (Result: $x = -1, L - R = 0$)

E. SEQUENCE OF NUMBERS

Arithmetic Sequence

Sequence of numbers can be generated using the calculator. To generate or form a particular sequence:

Punch the first term and press [=] add the common difference and press [=]. Keep on pressing the equal button [=] until the expected last term is achieved.

Example 1

Generate a sequence with 2 as the first term and the common difference 3.

Procedure

2 [=][+]3 [=](Ans: 5)[=](Ans: 8)[=](Ans: 11)[=](Ans: 14)

[=](Ans: 17)[=](Ans: 20)[=](Ans: 23)[=](Ans: 26)[=]

(Ans: 29) ...

Therefore, the sequence is generated as; 2, 5, 8, 11, 14, 17, 20, 23, 26, 29, ...

Example 2

List the first 15 terms of the sequence with 1 as the first term and 5 as the common difference.

Procedure

1 [=] [+] 5 [=] (ans: 6) [=] (ans: 11) [=] (ans: 16)

[=](ans: 21)[=](ans: 26)[=](ans: 31)[=] (ans: 36)

[=] (ans: 41) [=] (ans: 46)[=] (ans: 51) [=] (ans: 56)

[=] (ans: 61) (ans: 66) [=] (ans: 71) ...

Now the sequence is generated as; 1 ,6, 11, 16, 21, 26, 31, 36, 41, 46, 51, 56, 61, 66, 71, ...

Geometric Sequence

For geometric sequence, punch the first term and press [=], multiply the result by the common ratio and keep pressing [=] until the required last term is achieved.

Example 3

Write down the first ten terms of the geometric sequence with a first term of 1 and a common ratio of 3

Procedure

1 [=] [×] 3 [=] (*Ans*: 9) [=] (*Ans*: 27) [=] (*Ans*: 81)

[=](*Ans*: 243)[=](*Ans*: 729)[=](*Ans*: 2187)

[=] (*Ans*: 6561) [=] (*Ans*: 19683)

Hence the sequence is 1, 3, 9, 27, 81, 243, 729, 2187, 6561, 19683

Multiples Of A Number Could Also Be Generated Using The Following Procedure:

➢ Punch the number whose multiples are to be listed

➢ Press equal to [=]

➢ Add the number to the answer (i.e. ANS [+] the number) and keep pressing the [=]

Example 4

List the multiple of 3 between 1 and 60

Procedure

3 [=] (Ans: 3) [+] 3 [=] (Ans: 6)[=] (Ans: 9) [=] (Ans: 12) [=] (Ans: 15) [=] (Ans: 18) [=] (Ans: 21) [=] (Ans: 24) [=] (Ans: 27) [=] (Ans: 30) [=] (Ans: 33) [=] (Ans: 36) [=] (Ans: 39) [=]

(Ans:42) [=] (Ans: 45) [=] (Ans: 48) [=] (Ans: 51) [=] (Ans: 54) [=] (Ans: 57)

Now the multiple of 3 are: 3, 6, 9, 12, 15, 18, 21, 24, 27, 30, 33, 36, 39, 42, 45, 48, 51, 54 *and* 57

F. Summation

This allows you to calculate the sum of a series between two terms. The method for this essentially is the same as that for integration. To access the summation function, press [SHIFT] $\sum_0^{[]}[$]and cursor will be flashing for you to enter the expression (in terms of x), using ALPHA and then specify the first and the number of terms of the sequence. For arithmetic progression, the expression should be x [+] the common difference.

Get Your Own Question Now!

Hope You Enjoyed Solving Your Own Questions! Good Work!

NUMBER BASES

The number base can be called from the calculator using the following mode operation **[MODE] [4]**. The calculator can be used to calculate the following bases: **Denary/Decimals** (base ten), **Hexadecimal (Base Sixteen), Binary (Base Two)** and **Octary (Base 8)**. The bases are represented as BIN-Binary, DEC-Decimal, OCT-Octary and HEX-Hexadecimal. This allows you to add / subtract /divide and multiply in different number bases. The green labels are specialized for number base calculation.

Decimal mode acts much like the normal calculator mode; except you cannot enter fractions and it will not compute numbers that are too high or too low.

Octal mode is a number system in base 8. You can enter up to 11 octal digits in this mode

Hexadecimal mode allows you to enter up to 8 hexadecimal digits in a single term. Hexadecimal is shorthand for numbers represented in binary.

To Convert One Base to Another

Example 1

Convert 1001_2 to base ten.

Procedure

[MODE] [4] [BIN] [1001] [=] [DEC] (Result: 9). Thus the result is 9 in base ten.

Example 2

Convert 2354_{ten} to base sixteen.

Procedure

[MODE] [4] [DEC] [2354] [=] [HEX] (Result: 932_{16})

Example 3

Convert 226_{ten} to number base 8

Procedure

[MODE] [4] [DEC] [226] [=] [OCT]($Result$: 342_8)

Example 4

Arrange1101_2, 42_{16} and 28_{10} in descending order of magnitude.

Procedure

First, convert each to base ten and compare them

1101_2 = [MODE] [4] [BIN] [1102] [=] [DEC] (Result: 13)

$42_{16} =$ [MODE] [4] [HEX] [42] [=] [DEC] (Result: 66)

$28_{10} =$ 28

Now arrange them in descending order, we have

$42_{16}, \ 28_{10}, \ 1101_2$

Example 6

Convert 2617_{eight} to base ten numeral.

Procedure

[MODE] [4] [OCT] [2617] [=] [DEC] $(Result: 1423_{ten})$

Example 7

Evaluate in base two $100111_2/11_2$

Procedure

[MODE] [4] [BIN] [100111] [=] [÷] [11] [=] $(Result: 1101_2)$

Example 8

Simplify the following $1322_8 \times 13_8$

Procedure:

[MODE] [4] [OCT] [1322] [=] [×] [13] [=](Result: 17406_8)

To Perform Division Of Numbers In Bases Other Than Ten, It Is Advisable To First Convert The Number To Base

Ten, Perform The Division And Then Convert The Result Back To The Original Base. This Is Quite Tedious; Using The Calculator Makes It Easier.***

What Do You Expect? Do The Right Thing!

RELATIONS

<u>Completing Table for Relations</u>

Complete and complete the table of values for the relation $y = x^2 - x - 2$ for $-3 \leq x \leq 4$

x	-3	-2	-1	0	1	2	3	4
y	10				-2			10

Procedure

1. [MODE] [7]

2. $[x^\blacksquare]$ $[ALPHA]$ [ )] $[\rightarrow]$ [2] $[\rightarrow]$ $[-]$ $[ALPHA]$ [ )] $[-]$ 2 [=]

3. [Start?] , here you enter the lower limit for the interval given. In this question the lower limit is [-3] [=].

4. [End?], here you enter the upper limit for the interval given. In this question the upper limit is [4] [=] [=]

5. Table displays

x	-3	-2	-1	0	1	2	3	4
y	10	4	0	-2	-2	0	4	10

Example 2

Complete the table of values for the relation $y = 5x^2 + 2x - 3$

for the range $-4 \leq x \leq 4$

x	-4	-3	-2	-1	0	1	2	3	4
y					-3				

Procedure

1. [MODE] [7]
2. $[x^\blacksquare]$ [5] [ALPHA] [ )] [→] [2]
   [→] [+] [2] [ALPHA] [ )] [−] 3 [=]
3. [Start?], here you enter the lower limit for the interval given. In this question the lower limit is [−4] [=].
4. [End?], here you enter the upper limit for the interval given. In this question the upper limit is [4] [=] [=]
5. Table displays

x	-4	-3	-2	-1	0	1	2	3	4
y	69	36	13	0	-3	4	21	48	85

Example 3

Copy and complete the table of values of $y = x^3 + 2x^2 - 13x + 10$ for the range $-2 \leq x \leq 5$ below;

x	-2	-1	0	1	2	3	4	5
y			10					

Procedure

6. [MODE] [7]

7. $[x^{\blacksquare}]$ [ALPHA] [ )] [→] [3]

   [→] [+] [2] [ALPHA] [ )] $[x^{\blacksquare}]$ [2]

   [→] [−] [13] [ALPHA] [ )] [+] 10 [=]

8. [Start?], here you enter the lower limit for the interval given. In this question the lower limit is [−2] [=].

9. [End?], here you enter the upper limit for the interval given. In this question the upper limit is [5] [=] [=]

10. Table displays

x	-2	-1	0	1	2	3	4	5
y	36	24	10	0	0	16	54	120

Determination of Unknowns

Example 4

What is the value of $\frac{ABC}{AB-BC}$ if $A = -3, B = 2$ and $C = -2$

Procedure

Punch the relation given onto the calculator memory

1. $\left[\frac{\blacksquare}{\blacksquare}\right]$ $[ALPHA][A][ALPHA][B][ALPHA][C]\,[\rightarrow][ALPHA][A]$
   $[ALPHA][B][-][ALPHA][B][ALPHA][C]$

2. Press the [**CALC**] key to allocate the value of each variable into the calculator memory

3. [CALC] (A?) [(−)] 3 [=] (B?)2 [=](C?)[(−)] 2
   [=] (result: − 6)

Example 5

Given that $a = 2$ and $b = -3$, evaluate $a(a^2 + b^2)$

Procedure

1. [ALPHA] [A] [( ] $[x^{\blacksquare}]$ [ALPHA][A] $[\rightarrow]$ [2] $[\rightarrow]$ [+] $[x^{\blacksquare}]$
   [ALPHA][B] $[\rightarrow]$ [2] $[\rightarrow]$)

2. [CALC] (A?) 2 [=] (B?) [(−)] 3 [=]   (Result: 26)

To Determine a Value For An Unknown

Example 6

The time t seconds for one complete oscillation of a simple pendulum of length I meters is given by $t = 2\pi\sqrt{l/g}$ where gm^{-2} is the acceleration due to gravity. Find the length of pendulum if $t = 1s$ and $g = 9.8ms^{-2}$.

Procedure

$$t = 2\pi\sqrt{l/g}$$

NB: it can be noted that t, I and g are not found on the alphabet keys of the calculator. Therefore it is advisable to represent each by different letters. Let us represent t by M, l by X and g by Y.

1. $[ALPHA]$ $[M]$ $[ALPHA]$ $[=]$ 2 $[SHIFT]$ $[\pi]$ $[\sqrt{\blacksquare}]$ $\left[\frac{\blacksquare}{\blacksquare}\right]$ $[ALPHA]$ $[X]$ $[\rightarrow]$ $[ALPHA][Y]$ $[\rightarrow]$
2. $[SHIFT][SOLVE](M?)1 [=] (Y?)9.8 [=]$
   $[=]$ (Result: 0.25, $L - R = 0.2482368999$)

Example 7

Find the value of x in the equation below

$$4x + \frac{x-3}{x+10} = 10$$

Procedure

1. 4 $[ALPHA][\,)\,]$ $[+]$ $\boxed{\dfrac{\blacksquare}{\ }}$ $[ALPHA][\,)\,]$ $[-]$ 3 $[\rightarrow]$ $[ALPHA]$
   $[\,)\,]$ $[+]$ 10 $[\rightarrow]$ $[ALPHA]$ $[CALC]$ 10
2. $[SHIFT]$ $[SOLVE]$ $[=](Result: 2.51)$

TRY ME OUT!

1. Find R_1 in the relation below if $R_2 = 8$ and $R = 5$, $\dfrac{1}{R} =$
   $\dfrac{1}{R_1} + \dfrac{1}{R_2}$ (Result: $= \dfrac{40}{3}$)
2. If $3y = 2x^2 - 3x + 7$, find y when $x = 7$ (Result: 28)
3. Solve for the value of p if $2(p + 4) = 7p + 2$ (Result: $^6/_5$)
4. Solve $\dfrac{7x+3}{2} - \dfrac{9x-8}{4} = 6$ (result: $x = 2$)
5. Solve $\sqrt{x^2 + 33} = x + 3$ (result: $x = 4$)

NB: When solving series of such questions the memory must be cleared before tackling the next question. Thus [SHIFT] [9] [3] [=]

VECTORS

This mode allows you to perform calculations on **3D** and **2D** vectors - up to three at a time. Upon selecting vector mode, you'll be prompted to choose a vector memory slot to enter (VctA, VctB or VctC). After choosing which memory slot you're going to use, you will be prompted to choose the dimensions of the vectors (either 2 or 3). Now enter the values of your vector and then press [=] after entering the value in the each vector.

[MODE] [8] turns the vectors function on. Let us try to understand how it is being manipulated using the following examples.

Example 1

If $A = \begin{pmatrix} 3 \\ 4 \end{pmatrix}$ and $B = \begin{pmatrix} 2 \\ 3 \end{pmatrix}$. Find $2A + 3B$

Procedure

1. [MODE] [8]
2. [1] [2] [3] [=] 4 [=] {thus input of the first vector}
3. [SHIFT] [5] [1] [2] [2] 2 [=] 3 [=]
   {thus input of the second vector}
4. [ON] 2 [SHIFT][5][3] [+]3 [SHIFT][5][4] [=]
   [ANS] 12 [→] 17 {thus the result is $\begin{pmatrix} 12 \\ 17 \end{pmatrix}$

Manually, $2A + 3B = 2\begin{pmatrix}3\\4\end{pmatrix} + 3\begin{pmatrix}2\\3\end{pmatrix} = \begin{pmatrix}6\\8\end{pmatrix} + \begin{pmatrix}6\\9\end{pmatrix} = \begin{pmatrix}12\\17\end{pmatrix}$

NB: For vctA(m) m? The m represents the number column vector. In the example above m=2 since it is two column vector. The vctA, vctB and vctC represent the first, second and the third vector respectively.

Example 2

Given $X = \begin{pmatrix}1\\3\end{pmatrix}, Y = \begin{pmatrix}3\\5\end{pmatrix}$ and $Z = \begin{pmatrix}2\\4\end{pmatrix}$. Find $X + 5(Y + Z)$

Procedure

1. [MODE] [8]
2. [1] [2] [1] [=] 3 [=] {thus input of the first vector}
3. [SHIFT] [5] [1] [2] [2] 3 [=] 5 [=]
   {thus input of the second vector}
4. [SHIFT] [5] [1] [2] [3] 2 [=] 4 [=]
   {thus input of the third vector}
5. [ON] [SHIFT] [5] [3] [+] 5
   ([SHIFT][5] [4] [+] [SHIFT] [5] [5])[=]
   [ANS] 26 [→] 48 {thus the result is $\begin{pmatrix}26\\48\end{pmatrix}$

NB: note that the variable X,Y and Z has been represented by A,B and C respectively as specified in the vector mode.

Example 3

If $a = 5i + j - 3k$ and $b = 2i - 2j - 7k$, find $2a + b$

Procedure

1. MODE] [8]
2. [1] [1] [4] [=] 1 [=] − 3 [=]

   {thus input of the first vector}
3. [SHIFT] [5] [1] [2] [1] 2 [=] − 2 [=] − 7 [=]

   {thus input of the second vector}
4. [ON] 2 [SHIFT] [5] [3] [+] [SHIFT][5][4] [=]

   [ANS]12 [→] 0 [→] − 13   (Result: 12i + 0j − 13k)

Note that this is a 3D vector

Let Us Try Working It Manually To Verify:

$2a + b = 2(5i + j - 3k) + (2i - 2j - 7k)$

$= (10i + 2j - 6k) + (2i - 2j - 7k)$

$= (10 + 2)i + (2 + (-2)j + (-6 + (-7)k$

$= 12i - 13k$

THE DOT PRODUCT OF A VECTOR

Guess you have been wondering what option [SHIFT] [7] (dot) in vector option is used for. It is used for calculating the dot product of vectors.

To work out the dot product of a vector, select vector A (the first vector), re-enter vector options, select dot {[SHIFT] [1]}, go into vector options one more time and select vector B (the second vector). Then press [=].

Example 4

Find the scalar or dot product (a.b) when

    a. $a = i + 2j - k$ $b = 2i + 3j + k$

    b. $a = 2i + 3j + 4k$ $b = 5i - 2j + k$

Procedure

a.

1. MODE] [8]
2. [1] [1] [1] [=] 2 [=] – 1 [=]

    {thus input of the first vector}

3. [SHIFT] [5] [1] [2] [1] 2 [=] 3 [=] 1 [=]

    {thus input of the second vector}

4. [ON] [SHIFT] [5] [3] [SHIFT] [7] [SHIFT] [5] [4] [=]

    (Result: 7)

b.

1. MODE] [8]
2. [1] [1] [2] [=] 3 [=] 4 [=] {thus input of the first vector}
3. [SHIFT] [5] [1] [2] [1] 5 [=] – 2 [=] 1 [=]

    {thus input of the second vector}

4. [ON] [SHIFT] [5] [3] [SHIFT] [7] [SHIFT] [5] [4] [=]
    (Result: 8)

MAGNITUDE OF VECTORS

To determine the magnitude of a vector, the following procedure is used:

Press [SHIFT] [Abs] [go into vector options and choose the vector] and press [=].

Example 5

Find the magnitude $A = \begin{pmatrix} 3 \\ 4 \end{pmatrix}$

Procedure

1. MODE] [8]
2. [1] [2] [3] [=] 4 [=]
3. [ON][SHIFT] [Abs] [SHIFT] [5] [3]) [=] (Result: 5)

Example 6

Determine the length of $A = \begin{pmatrix} 3 \\ 5 \end{pmatrix}$.

Procedure

1. MODE] [8]

2. [1] [2] [3] [=] 5 [=]

3. [ON][SHIFT] [Abs] [SHIFT] [5] [3]) [=] (Result: 5.831)

Example 7

Given that $P = \binom{6}{4}$ and $Q = \binom{3}{5}$, find the value of $|4P + Q|$

Procedure

1. [MODE] [8]

2. [1] [2] 6 [=] 4 [=] {thus input of the first vector}

3. [SHIFT] [5] [1] [2] [2] 3 [=] 5 [=] {thus input of the second vect

4. [ON] [SHIFT] [Abs] [SHIFT] [5] [3] [+][SHIFT] [5] [4]) [=
    ] (Result: 12.728)

Cut The Joke! Practice More Examples

Now!

STATISTICAL FUNCTION

Statistics mode allows inputting data in multiple forms which gives you the procedures required to calculate many values from the data sets (e.g. Frequency, standard deviation, etc.) it requires the 3D-mode: **[MODE] [3]**

All input should be started **[SHIFT] [9] [3] [=] [=]** with to clear the statistical memory. The statistical function key ([SUM] and [VAR]) is mainly used to compute the results.

The following examples will help you to understand how statistical problems are solved using the calculator.

General Procedure

1. Activate the statistical mode by pressing [MODE] [3]
2. Input each data followed by [=]. The statistical memory should be cleared before the data is inputted.
3. Use the [SHIFT] [STAT] to select the operation you want the calculator to perform and press [=].

Example 1

Find the mean of the following data: 8, 3, 5, 2, 3, 5, 6, 1, 1, 2

Procedure

1. [SHIFT] [9] [3] [=] [=]
2. [MODE] [3] [1]
3. 8 [=] 3 [=] 5 [=] 2 [=] 3[=] 5 [=] 6[=] 1 [=] 1 [=] 2 [=]
4. [ON] [SHIFT] [STAT]
5. Mean: [4] [2] [=](result: 3.6)

Don't Be Surprise! Work It Out and Check!

Example 2

The following are the marks obtained by students in an examination:

Marks	5	6	7	8	9	10
No. of students	3	2	11	17	6	1

Calculate

    a. The mean mark

    b. Standard deviation

    c. Total number of students (frequency)

Procedure

1. [SHIFT] [9] [3] [=] [=]
2. [MODE] [3] [1]
3. [SHIFT] [MODE] [↓] [4] [1]
4. Enter the marks in the column marked (x)
5. 5 [=] 6 [=] 7 [=] 8 [=] 9 [=] 10 [=]
6. Enter the frequency in the column marked (**FREQ**) by using the center key to move the cursor to the right [→] and up [↓]. Note the marks **SHOULD** correspond to the frequency.
7. 3 [=] 2 [=] 11 [=] 17 [=] 6 [=] 1 [=]
8. [ON] [SHIFT] [STAT] Mean: [4] [2] [=](result: 7.6)
9. [ON] [SHIFT] [STAT] Standard deviation: [4] [3] [=](result: 1.1135)
10. [ON] [SHIFT] [STAT] Total number of students (frequency): [4] [1] [=] (result: 40)

Example 3

Calculate the standard deviation from the following table:

Age (yrs)	6	7	8	9	10	11
Frequency	4	6	10	11	8	1

Procedure

1. [SHIFT] [9] [3] [=] [=]
2. [MODE] [3] [1]
3. [SHIFT] [MODE] [↓] [4] [1]
4. Enter the marks in the column marked (x)
5. 6 [=] 7 [=] 8 [=] 9 [=] 10 [=] 11 [=]
6. Enter the frequency in the column marked (**FREQ**) by using the center key to move the cursor to the right [→] and up [↓]. Note the marks **SHOULD** correspond to the frequency.
7. 4 [=] 6 [=] 10 [=] 11 [=] 8 [=] 1 [=]
8. [ON] [SHIFT] [STAT] Standard deviation: [4] [3] [=] (result: 1.3)

Example 4

The score obtained by students in a test are 21, 25, 27, 25, 27, 21, 24, 23, 23 and 24. Calculate

1. The mean score
2. The standard deviation
3. Variance

Procedure

1. [SHIFT] [9] [3] [=] [=]
2. [MODE] [3] [1]
3. 21 [=] 25 [=] 27 [=] 25 [=] 27[=] 21 [=] 24 [=] 23 [=] 23 [=] 24 [=]
4. [ON] [SHIFT] [STAT] Mean: [4] [2] [=](result: 24)
5. [ON] [SHIFT] [STAT]   Standard deviation: [4] [3] [=] (result: 2)
6. [ON] [SHIFT] [STAT] Variance: [4] [4] [=] (result: 4)

Example 5

Twenty students score the following marks in a chemistry test:

5, 2, 5, 3, 1, 6, 2, 2, 3, 4, 2, 1, 2, 2, 4, 3, 2, 2, 2, 3. Find the

a. Mean mark scored
b. Population standard deviation

Procedure

1. [SHIFT] [9] [3] [=] [=]
2. [MODE] [3] [1]

3. 5 [=] 2 [=] 5 [=] 3 [=] 1 [=] 6 [=] 2 [=] 2 [=] 3 [=] 4
   [=] 2 [=] 1 [=] 2 [=] 2 [=] 4
   [=] 3 [=] 2 [=] 2 [=] 2 [=] 3 [=]

4. [ON] [SHIFT] [STAT] Mean: [4] [2] [=](result: 2.8)

5. [ON] [SHIFT] [STAT]   Standard deviation: [4] [3] [=] (result: 1.33)

Try me out!

The biology scores of ten students in a class are as follows:

Biology	30	50	70	73	26	60	42	38	92	49

   a. Find the mean of the scores in biology.

   b. Find the standard deviation of the scores in biology.

   Results

   a. Mean score in biology: 53

   b. Standard deviation of scores in biology: 19.72

SUMMARY

To Recall This Type Of Value	Key Sequence	Description
n	[SHIFT] [STAT] [4] [1]	No. of data (frequency)
$\sum x$	[SHIFT] [STAT] [3] [2]	Sum of value
$\sum x^2$	[SHIFT] [STAT] [3] [1]	Sum of squares of values
$\breve{x}$	[SHIFT] [STAT] [4] [2]	Mean
$x\sigma n$	[SHIFT] [STAT] [4] [3]	Standard deviation

While inputting data or after inputting data is complete, you can use the [↑] and [↓] to scroll through data you have input. You can then edit the displayed data if you want. Input the new value and then press the [=] key to replace the old value with the new one.

Regression Calculations

Press [MODE] [3] ([2] or [3] or [4] or [5] etc) to launch the regression mode depending on the type of regression you want to perform (i.e. linear, exponential, logarithm, quadratic function, etc.). This mode allows you to find the mean of x of y, deviation of x and y, gradient (B) and intercept on y axis (A).

Example 1

The following data is regressed with the least square regression to $y = a_0 + a_1 x$. Find the value a_1 (gradient of the graph).

x	1	20	30	40
y	1	400	800	1300

Procedure

1. [SHIFT] [9] [3] [=] [=]
2. [MODE] [3] [2]
3. Enter the marks for x in the column marked x
4. 1 [=] 20 [=] 30 [=] 40 [=]
5. Enter the marks for y in the column marked **y** by using the center key to move the cursor to the right [→] and up [↓].
6. 1 [=] 400 [=] 800 [=] 1300 [=]
7. [ON] [SHIFT] [STAT] [5] [2] [=](result: 32.625)

Example 2

The data in the table below shows the year's experiences and salary of seven workers selected at random in an industry.

Workers	A	B	C	D	E	F	G
Years of experience (x)	3	7	4	9	11	16	8
Salary GH¢ (,000) (y)	13	15	14	17	15	19	17

Find

a. $\check{x}$ the mean of x

b. $\bar{y}$ the mean of y

Procedure

1. [SHIFT] [9] [3] [=] [=]
2. [MODE] [3] [2]
3. Enter the years of experience in the column marked x
4. 3 [=] 7 [=] 4 [=] 9 [=] 11 [=] 16 [=] 8 [=]
5. Enter the salary in the column marked **y** by using the center key to move the cursor to the right [→] and up [↓].
6. 13 [=] 15 [=] 14 [=] 17 [=] 15 [=] 19 [=] 17 [=]
7. [ON] [SHIFT] [STAT] Mean of x: [4] [2] [=] (result: 8.3)
8. [ON] [SHIFT] [STAT] Mean of y: [4] [5] [=] (result: 15.7)

Example 3

The table below the value of two variables x and y obtained from a survey.

x	11	12	14	15	16	18	19	20
y	81	73	53	53	43	29	15	9

Find

   a. $\check{x}$ and $\check{y}$ the means of x and y respectively

   b. The equation of the line

Procedure

1. [SHIFT] [9] [3] [=] [=]
2. [MODE] [3] [2]
3. Enter the marks x in the column marked x
4. 11 [=] 12 [=] 14 [=] 15 [=] 16 [=] 18 [=] 19 [=] 20 [=]
5. Enter the marks y in the column marked **y** by using the center key to move the cursor to the right [→] and up [↓].
6. 81 [=] 73 [=] 53 [=] 53 [=] 43 [=] 29 [=] 15 [=] 9 [=]
7. [ON]   [SHIFT]   [STAT]   Mean   of x: [4]   [2] [=] (result: 15.625)
8. [ON] [SHIFT] [STAT] Mean of y: [4] [5] [=] (result: 44.5)

NB: The general equation of a linear graph used by the calculator is given as $y = A + Bx$
Where A and B are the intercept on y-axis and the gradient of the line respectively.

9. [ON]   [SHIFT]   [STAT]   [5]   [1] [=]   (result: Intercept 167.91)

10. [ON] [SHIFT] [STAT] [5] [2] [=] (result: Gradient –

7.90)

From the results, the equation of the line is:

$y = 167.91 - 7.90x$

 *Try Me Out*

The table below shows the mark of 10 students in physics and

mathematics test.

Math (x)	75	80	93	65	87	71	98	68	84	77
Physics (y)	82	78	86	72	91	80	95	72	89	74

Find

(a) $\check{x}$ and $\check{y}$, the mean of x and y respectively

(b) The equation of the line of best fit

Results

(Intercept: 29.13) , (Gradient: 0.66) ,Now the equation of the

straight line is: $y = 0.66x + 29.13$

***This procedure can also be applied on logarithm,

exponential, power inverse and quadratic functions. ***

CALCULUS

DIFFERENTIATION AND INTEGRATION

NUMERICAL DIFFERENTIATION

Your fx - 991ES/Plus scientific calculator makes the solving of differential of quantities easier. It allows you to punch the expression as it appears on the question paper. The calculator will, given $f(x)$, calculate $f^1(x)$ for a particular value of x. Press **[SHIFT]** $[\frac{d}{dx}]$ this will bring up something that looks a bit like $\frac{d}{dx}[\blacksquare]\big|x = [\blacksquare]$. Enter $f(x)$, the function to be differentiated and scroll right to enter the value of x at which you want to calculate the gradient. Finally press [=] for the result. The numerical differentiation operation **[SHIFT]** $[\frac{d}{dx}]$ takes 2 arguments:

1. The function of x to differentiate
2. The point where the derivative is evaluated at

General procedure

1. Press the **[SHIFT]** $[\frac{d}{dx}]$ key and punch the relation to be differentiated.

2. Press [→] followed by the point or value at which the derivative is differentiated.

3. Press the [=] for the result. The result might take some few seconds to be evaluated.

Example 1

If $y = x^4 + 6x^2 + 9$. Find the gradient at $x = \frac{1}{2}$

Procedure

[SHIFT] $[\frac{d}{dx}]$

$[x^\blacksquare]$ [ALPHA] [ )] [→] 4 [→] [+] 6 [ALPHA] [ )] $[x^\blacksquare]$

[2] [→] [+] 9 [→] [0.5] [=] (Result: 6.5)

Example 2

Find the gradient of the curve $y = (x - 3)(x^2 + 2)$ at the point where $x = 1$

Procedure

[SHIFT] $[\frac{d}{dx}]$

([ALPHA] [ )] [−] [3]) ($[x^\blacksquare]$[ALPHA] [ )] [→] 2 [→] [+] 2)

[→] [1] [=]   (Result: -1)

Example 3

Find the gradient of the curve $y = x^3 + 7x - 4$ at the point where $x = 3$

Procedure

SHIFT] $[\frac{d}{dx}]$

$[x^\blacksquare]$ [ALPHA] [ )] [→] 3 [→] [+] 7 [ALPHA] [ )] [−] 4

[→] [3] [=] (Result: 34)

So Easy! Try These Out!

1. Find the gradient of $y = \frac{3x-1}{x}$ at the point where $x = \frac{1}{2}$

   (*Result*: 4)

2. Given that $y = \cos(7x + 2)$, find $\frac{dy}{dx}$ at $x = 4$ (*Result*: 3.5)

3. Suppose that the amount of water in a holding tank at t minutes is given by $V(t) = 2t^2 - 16t + 35$. Determine the volume of water at $t = 5$. (*Result*: 4)

INTEGRATION

The fx - 991ES/Plus makes the solving of definite integral questions straight forward: it allows you to punch the expression as it appears on the question paper.

To perform definite integration, first press the $\int_a^b dx$ button. This will bring up an integral sign with blanks for the upper and lower limits and the integrand and a dx at the end. The cursor will be flashing for you to enter the integrand, (in terms of x), using alpha. For example $\int_1^2(2x^2 + 6)dx$. Once you have entered your integrand $2x^2 + 6$, input your upper and lower limits by scrolling up and down using the replay or directional arrow buttons. Press [=] and it will perform the integration, giving the answer as a fraction if appropriate.

Now let us illustrate what has been discussed with examples:

Example 1

Evaluate the integral $\int_1^2(x^2 + 3)dx$

Procedure

$[\int dx]$ $[x^\blacksquare]$ $[ALPHA]$ $[\,)\,]$ $[\rightarrow]$ 2 $[\rightarrow]$ $[+]$ 3 $[\downarrow]$ 1 $[\uparrow]$ 2 $[\rightarrow]$ $[=]$

$(Result: 5.333333 = \frac{16}{3})$

Example 2

Find the value of the integral $\int_2^3 (x^2 + 2x - 1)dx$

Procedure

$\left[\int dx\right]$ $[x^\blacksquare]$ $[ALPHA]$ $[\,)\,]$ $[\rightarrow]$ 2 $[\rightarrow]$ $[+]$ 2 $[ALPHA]$ $[\,)\,]$ $[-]$

1 $[\downarrow]$ 2 $[\uparrow]$ 3 $[\rightarrow]$ $[=]$ (Result: $10.3333 = \frac{31}{3}$)

Example 3

Determine the exact integral of $\int_0^{\frac{4}{\pi}} 2\cos 2x dx$.

Procedure

$[\int dx]$ 2 $[\cos]$ 2 $[ALPHA]$ $[\,)\,]$

$[\downarrow]$ 0 $[\uparrow]$ $\left[\frac{\blacksquare}{\blacksquare}\right]$ $[SHIFT]$ $[\pi]$ $[\rightarrow]$ 4 $[\rightarrow]$ $[=]$ (result: 1.6)

1. Evaluate $\int_0^2 (x+1)\sqrt{x^2 + 2x}\, dx$ (Result: 7.542)

2. Determine the value of $\int_{-2}^5 (x^2 - 3x - 10)dx$ (Result: - 57.16667$=\frac{-343}{6}$)

3. Evaluate the integral $\int_0^{\frac{4}{2}} sin^2 x cos x \ dx \quad \left(Result: 0.2 = \frac{1}{5} \right)$

4. Find the area under the curve $f(x) = \frac{300x}{1+e^x}$. From $x =$ $0 \ to \ x = 10 \ (result: 246.59)$

PERMUTATION AND COMBINATION

The permutation operation (nPr) and the combination operation (nCr) is found on the multiplication (×) and the division sign (÷) respectively.

When dealing with permutation or combination, no mode is required.

Example 1

If a club has 20 members, how many different four members committee are possible?

Procedure

The above question is represented by $20C_4 = 4845$

20 [SHIFT] [÷] 4 [=] (result: 4845)

Example 2

Evaluate the following:

    a. $9C_6$

    b. $7C_4$

    c. $8P_5$

    d. $10P_3$

    e. $15C_7$

f. $4P_4$

Procedure

a.   9 [SHIFT] [÷] 6 [=] (Result: 84)

b.   7 [SHIFT] [÷] 4 [=] (Result: 35)

c.   8 [SHIFT] [×] 5   [=] (Result: 6720)

d.   10 [SHIFT] [×] 3 [=] (Result: 720)

e.   15 [SHIFT] [×]   7 [=] (Result: 6435)

f.   4   [SHIFT] [×] 4 [=] (Result: 24)

MATRICES

The matrix function (MAT) must first be activated. The procedure is similar to that of the vector. The dimensions of the matrix would be required thus the number of rows and the columns of the matrix.

General Procedure

1. Activate the matrix mode by pressing : **[MODE] [6]**
2. Enter the values of the matrix onto the calculator memory. You would be asked to enter the types of the matrix or the dimension of the matrix. Thus for MatA (m × n), the m is for number of rows and the n is for number of columns. To do this, for example, with a 3 × 2 matrix you would press [2].
3. Now choose the operation that you want to do with the inputted data. For example could be determinant, transpose of the matrix etc.
4. Use the arrow keys to extract or move through the answer.

NB: The general form of the matrix is shown on the calculator as:

1. 3 × 3
2. 3 × 2
3. 3 × 1
4. 2 × 3

5. 2 × 2

6. 2 × 1

    [↓]

1. 1 × 3

2. 1 × 2

3. 1 × 1

Example 1

If $A = \begin{pmatrix} 1 & 2 \\ 3 & 4 \end{pmatrix}$ and $B = \begin{pmatrix} -1 & 2 \\ -3 & 1 \end{pmatrix}$ find AB.

Procedure

1. [MODE] [6]
2. To enter matrix A: [1] [5] 1 [=] 2 [=] 3 [=] 4 [=]
3. To enter matrix B: [SHIFT][MAT][2][2][5] [(−)]1
    [=] 2 [=] [(−)] 3 [=] 1 [=]
4. [ON] [SHIFT] [MATRIX] [3] [×] [SHIFT] [MATRIX]
    [4] [=] Result: $\begin{pmatrix} -7 & 4 \\ -15 & 10 \end{pmatrix}$

Example 2

Given that A $= \begin{pmatrix} 2 & 1 \\ 3 & 4 \end{pmatrix}$, find the determinant of A.

Procedure

1. [MODE] [6]
2. [1] [5] 2 [=] 1 [=] 3 [=] 4 [=]
3. [ON] [SHIFT] [MATRIX] [7]
4. [SHIFT] [MATRIX] [3] [=] (Result: 5)

Example 3

A matrix of a linear transformation A is given $A = \begin{pmatrix} 2 & -5 \\ 1 & -3 \end{pmatrix}$, find the image of the point $(2, -4)$.

Procedure

1. [MODE] [6]
2. [1] [5] 2 [=] − 5 [=] 1 [=] − 3 [=]
3. [SHIFT] [MAT] [2] [2] [6] 2 [=] − 4 [=]
4. [ON] [SHIFT] [MATRIX] [3] [×] [SHIFT] [MATRIX] [4] [=] Result: $\begin{pmatrix} 24 \\ 14 \end{pmatrix}$

Example 4

If $A = \begin{pmatrix} 1 & -1 \\ 2 & 4 \end{pmatrix}$, find

a. A^{-1}
b. AA^{-1}

Procedure

1. MODE] [6]
2. [1] [5] 1 [=] − 1 [=] 2 [=] 4 [=]
3. [ON] [SHIFT] [MATRIX] [3] $[x^{-1}]$ [=]   Results:

$$\begin{pmatrix} 0.6667 & 0.1667 \\ -0.3333 & 0.1667 \end{pmatrix}$$

4. [ON] [SHIFT] [MATRIX] [3] [×] [SHIFT] [MATRIX] [3]

   $[x^{-1}]$ [=] (Result: $AA^{-1} = \begin{pmatrix} 1 & 0 \\ 0 & 1 \end{pmatrix}$)

Example 5

Find the transpose of $A = \begin{pmatrix} 1 & 2 \\ 3 & 4 \end{pmatrix}$

Procedure

1. [MODE] [6]
2. [1] [5] 1 [=] 2 [=] 3 [=] 4 [=]
3. [ON] [SHIFT] [MATRIX] [8]
4. [SHIFT] [MATRIX] [3] [=] (Result: $\begin{pmatrix} 1 & 3 \\ 2 & 4 \end{pmatrix}$

What Else? You Want More?

COMPLEX NUMBERS

In complex mode, the calculator is able to perform all expected calculations involving complex numbers. The **ENG** button (left of the open bracket button) becomes i. It is therefore possible to enter complex numbers in the form $a + bi$. Another option is to enter complex numbers in the modulus – argument form.

General Procedure

1. The complex number function must be on thus **[MODE] [2]**
2. Punch the complex relation given onto the calculator screen.
****Note that the imaginary part can be punched from [ENG]***
3. Enter the operation you want the calculator to perform and press [=]

Example 1

If $Z_1 = 3 + 2i$ and $Z_2 = -1 + i$. Find

a. $Z_1 + Z_2$
b. $Z_1 - Z_2$
c. $Z_1 Z_2$

Procedure

1. **[MODE] [2]**
2. $3 [+] 2 [ENG] [+] [(-)] 1 [+] [ENG] [=]$ *Result:* $2 + 3i$
3. $3 [+]2 [ENG] [-] [(-)] 1 [+] [ENG] [=]$ *Result:* $4 + 3i$
4. $3 [+] 2 [ENG] [×] [(-)] [+] [ENG] [=]$
   (*Result: thus* $3 - i$

Example 2

Find the real and imaginary parts of $\frac{2+3i}{3+2i}$

Procedure

1. $\left[\dfrac{\blacksquare}{\blacksquare}\right] 2 [+]3 [ENG][\rightarrow] 3 [+]2 [ENG] [=]$

   (*Result:* $\frac{12}{13} + \frac{5}{13}i$. The real part is $\frac{12}{13}$ and the imaginary part is $\frac{5}{13}i$

Example 3

Find the result of the following: $(3 + 4i)(2 - 5i)(1 - 2i)$

Procedure

1. $[(] 3 [+] 4 [ENG] [ )] [(] 2 [-] 5 [ENG] [)] [(] 1 [-] 2 [ENG] [ )] [=]$ Result: $12 - 59i$

Absolute Value and Conjugate Calculation

Supposing the imaginary number expressed by the by the rectangle form $z = a + bi$ is represented as a point in the Gaussian plane, you can determine the absolute value (r) and the argument (θ) of the complex number. The polar form is r $< \theta$. The unit of the argument of the angle is in degree. To obtain the absolute value (r) and the argument (θ), enter **[MODE] [2] [SHIFT] [MODE] [↓] [3] [2].**

Example 4

Express 3 + 4i in a conjugate form.

Procedure

1. [MODE] [2] [SHIFT] [MODE] [↓] [3] [2]
2. 3 [+] 4 [ENG] [=]   Result: 5 ∠ 53.1301°)

Example 5

Express 2 -5i in a conjugate form

Procedure

1. [MODE] [2] [SHIFT] [MODE] [↓] [3] [2]
2. 2 [−] 5 [ENG] [=]   Result: $\sqrt{29}$ ∠ −68.1986°

SCIENTIFIC CONSTANT

By pressing **[SHIFT]** **[CONST]** in any mode except base, it is possible to recall one of 40 constant stored on the calculator by inputting a number from 01 to 40 although none are necessary for a level exams or WASSCE (they will be given to you), the following may be helpful (be aware that examinations may except you to use specified rounded values)

NAME OF CONSTANT	SYMBOL	KEY SEQUENCE
Proton rest mass	m_p	[SHIFT] [CONST] [0] [1]
Neutron rest mass	m_n	[SHIFT] [CONST] [0] [2]
Electron rest mass	m_e	[SHIFT] [CONST] [0] [3]
Muon mass	m_μ	[SHIFT] [CONST] [0] [4]
Bohr's radius	$a_\circ$	[SHIFT] [CONST] [0] [5]
Planck constant	h	[SHIFT] [CONST] [0] [6]
Nuclear magneton	μ_N	[SHIFT] [CONST] [0] [7]

Bohr's magneton	μ_B	[SHIFT] [CONST] [0] [8]
Reduced planck constant	h	[SHIFT] [CONST] [0] [9]
Fine structure constant	α	[SHIFT] [CONST] [1] [0]
Classical electron radius	r_e	[SHIFT] [CONST] [1] [1]
Compton's wavelength	γ_c	[SHIFT] [CONST] [1] [2]
Proton gyromagnetic ratio	γp	[SHIFT] [CONST] [1] [3]
Proton Compton wavelength	γcp	[SHIFT] [CONST] [1] [4]
Neutron Compton wavelength	γcn	[SHIFT] [CONST] [1] [5]
Rydberg's constant	$R\infty$	[SHIFT] [CONST] [1] [6]

Atomic mass unit	u	[SHIFT] [CONST] [1] [7]
Proton magnetic moment	μp	[SHIFT] [CONST] [1] [8]
Electron magnetic moment	μe	[SHIFT] [CONST] [1] [9]
Neutron magnetic moment	μn	[SHIFT] [CONST] [2] [0]
Muon magnetic moment	$\mu \mu$	[SHIFT] [CONST] [2] [1]
Faraday's constant	F	[SHIFT] [CONST] [2] [2]
Electron charge	e	[SHIFT] [CONST] [2] [3]
Avogadro number	N_A	[SHIFT] [CONST] [2] [4]
Boltzmann constant	k	[SHIFT] [CONST] [2] [5]
Molar volume	V_m	[SHIFT] [CONST] [2] [6]

Universal gas constant	R	[SHIFT] [CONST] [2] [7]
Speed of light	$C_\circ$	[SHIFT] [CONST] [2] [8]
First radiation constant	C_1	[SHIFT] [CONST] [2] [9]
Second radiation constant	C_2	[SHIFT] [CONST] [3] [0]
Stefan-boltzmann constant	σ	[SHIFT] [CONST] [3] [1]
Permittivity of free space	$\varepsilon_\circ$	[SHIFT] [CONST] [3] [2]
Permeability of free space	$\mu_\circ$	[SHIFT] [CONST] [3] [3]
Magnetic flux quantum	$\emptyset_\circ$	[SHIFT] [CONST] [3] [4]
Acceleration due to gravity	g	[SHIFT] [CONST] [3] [5]
Conductance quantum	G	[SHIFT] [CONST] [3] [6]

Impedance of quantum	Z	[SHIFT] [CONST] [3] [7]
Difference between kelvin and Celsius	t	[SHIFT] [CONST] [3] [8]
Gravitational constant	G	[SHIFT] [CONST] [3] [9]
Atmospheric pressure	atm	[SHIFT] [CONST] [4] [0]

Example

A 50 kg person and a 75kg person are sitting on a bench. What is the magnitude of gravitational force each exerts on the other if they are 0.5m apart?

Procedure

$m_1 = 50kg; \ m_2 = 75kg; \ r = 0.5m$

$F = \dfrac{Gm_1m_2}{r^2} = \dfrac{G \times 50 \times 75}{0.5^2}$

Where G is the universal gravitational constant

$\left[\dfrac{\blacksquare}{\quad}\right]$ [SHIFT][$CONST$][39][×]50 [×]75 [→] 0.5 [x^2]

[=]   Result: 1.001142×10^{-6}

UNITS CONVERSION

Your calculator can convert between different units of measurement. The calculator's built-in conversion commands make it simple to convert values from one unit to another. You use the conversion commands in any calculation mode except for number bases. To recall a conversion command, press **[SHIFT]** and **[CONV]** which displays the conversion command menu and input the two digit number that corresponds to the conversion you want to recall. A full list of conversions is given below.

FROM	TO	KEY SEQUENCE
in	cm	[SHIFT] [CONV] [0] [1]
cm	in	[SHIFT] [CONV] [0] [2]
ft	m	[SHIFT] [CONV] [0] [3]
m	ft	[SHIFT] [CONV] [0] [4]
yd	m	[SHIFT] [CONV] [0] [5]
m	yd	[SHIFT] [CONV] [0] [6]
Mile	km	[SHIFT] [CONV] [0] [7]
km	mile	[SHIFT] [CONV] [0] [8]
n mile	m	[SHIFT] [CONV] [0] [9]
m	n mile	[SHIFT] [CONV] [1] [0]
acre	(m^2)	[SHIFT] [CONV] [1] [1]

m^2	acre	[SHIFT] [CONV] [1] [2]
gal(US)	l	[SHIFT] [CONV] [1] [3]
l	gal(US)	[SHIFT] [CONV] [1] [4]
gal(UK)	l	[SHIFT] [CONV] [1] [5]
l	gal(UK)	[SHIFT] [CONV] [1] [6]
pc	km	[SHIFT] [CONV] [1] [7]
km	pc	[SHIFT] [CONV] [1] [8]
km/h	m/s	[SHIFT] [CONV] [1] [9]
m/s	km/h	[SHIFT] [CONV] [2] [0]
oz	g	[SHIFT] [CONV] [2] [1]
g	oz	[SHIFT] [CONV] [2] [2]
lb	kg	[SHIFT] [CONV] [2] [3]
kg	lb	[SHIFT] [CONV] [2] [4]
atm	Pa	[SHIFT] [CONV] [2] [5]
Pa	atm	[SHIFT] [CONV] [2] [6]
mmHg	Pa	[SHIFT] [CONV] [2] [7]
Pa	mmHg	[SHIFT] [CONV] [2] [8]
hp	kw	[SHIFT] [CONV] [2] [9]
kw	hp	[SHIFT] [CONV] [3] [0]
kgf/cm^2	Pa	[SHIFT] [CONV] [3] [1]
Pa	kgf/cm^2	[SHIFT] [CONV] [3] [2]
kgf.m	J	[SHIFT] [CONV] [3] [3]

J	kgf.m	[SHIFT] [CONV] [3] [4]
lbf/in^2	kPa	[SHIFT] [CONV] [3] [5]
kPa	lbf/in^2	[SHIFT] [CONV] [3] [6]
°F	°C	[SHIFT] [CONV] [3] [7]
°C	°F	[SHIFT] [CONV] [3] [8]
J	cal	[SHIFT] [CONV] [3] [9]
cal	J	[SHIFT] [CONV] [4] [0]

Example 1

Change

    a.   10ft to meters(m)

    b.   1 inch to centimeter(cm)

Procedure

    a.   10 [SHIFT][CONV]03 [=](Result: 3.048m)

    b.   1 [SHIFT] [CONV] 01 [=] (Result: 2.54cm)

Example 2

Convert

    a.   1 atm of pressure to pascal

    b.   321000 pascal to atm

Procedure

a.   1 [SHIFT][CONV]25 [=](Result: 101325 Pa)

b.   321000 [SHIFT] [CONV] 26 [=] (Result: 3.168 atm)